EARTH FILES

ISLANDS

EARTH FILES – ISLANDS
was produced by

David West 👫 **Children's Books**
7 Princeton Court
55 Felsham Road
London SW15 1AZ

Editor: James Pickering
Picture Research: Carrie Haines

First published in Great Britain in 2002 by
Heinemann Library, Halley Court, Jordan Hill,
Oxford OX2 8EJ, a division of Reed Educational and
Professional Publishing Limited.

OXFORD MELBOURNE AUCKLAND
JOHANNESBURG BLANTYRE GABORONE
IBADAN PORTSMOUTH (NH) USA CHICAGO

06 05 04 03 02
10 9 8 7 6 5 4 3 2 1

ISBN 0 431 15620 4 (HB)
ISBN 0 431 15627 1 (PB)

British Library Cataloguing in Publication Data

Oxlade, Chris
Islands. - (Earth Files)
1. Islands - Juvenile literature
2. Island ecology - Juvenile literature
I. Title
551.4'2

PHOTO CREDITS :
Abbreviations: t-top, m-middle, b-bottom, r-right,
l-left, c-centre.

Front cover, 3, 4t, 5b, 8 both, 8-9t, 10-11t, 12tr, 12-
13, 16-17t, 22br, 25tl, 26r, 26-27, 27tr - Corbis
Images. 5tr (Gardar Igaliko), 6tl (Louise Murray),
11br, 20tr, 20-21m (Tony Waltham), 13tr, 19br
(Gavin Hellier), 14-15t (Nakamura), 16tr (Robert
Harding), 25r (Martyn F. Chillmaid), 26bl (C.
Bowman), 27tm (S. Grandadam), 28ml (K. Gillham),
4b & 14br, 11ml, 17tr, 27br - Robert Harding Picture
Library. 9mr, 18-19b, 20-21b (Peter Steyn), 13br
(McDougal Tiger Tops), 14-15b, 28-29 (Kurt
Amsler), 18-19t, 29tl (M. Watson), 21br (E.
Mickleburgh), 22tr (John Mason), 22ml (Adrian
Warren), 23r (D. Parer & E. Parer-Cook), 24bl
(François Gohier), 29bl (Jean-Paul Ferrero) - Ardea
London Ltd. 17br (Stephen Coyne) - Papilio. 18bl,
29mr - Popperfoto. 21tr - Werner Forman Archive.
23bl (Sally Morgan), 25bl (Edward Bent) - Ecoscene.
24bm - Mary Evans Picture Library.

Printed and bound in Italy

*An explanation of difficult words can be
found in the glossary on page 31.*

EARTH FILES

ISLANDS

Chris Oxlade

Heinemann
LIBRARY

CONTENTS

About half of the people on the huge island of New Guinea farm on the slopes of its high mountains.

Tropical islands are surrounded by spectacular coral reefs which provide a home for colourful fish and a wealth of other sea creatures.

INTRODUCTION

High, jagged, covered in jungle and surrounded by thousands of kilometres of sea, or low, flat, grassy, and off the coast of mainlands, the islands of the world are special places. The surrounding sea isolates them, creating worlds where unique plants and animals live, where the way people live has remained unchanged for thousands of years and where there are plentiful resources.

Greenland, the world's largest island, is mostly covered in ice. There are a few dozen settlements on the coast, where the snow melts in summer.

The Hawaiian Islands in the Pacific are formed by the summits of vast underwater volcanoes, some of which still spew out lava.

The Pacific Ocean contains thousands of coral atolls like this one in the Solomon Islands.

There are many thousands of islands dotted around the world. They range from rugged, mountainous islands in the icy polar regions, to tiny coral atolls scattered about the warm tropical oceans near the Equator.

Servenaya Zemiya

ARCTIC

New Siberian Islands

ASIA

Sakhalin

Aleutian Islands

Hokkaido

Honshu

Taiwan

Hainan

Hawaiian Islands

PACIFIC

Philippines

Borneo

New Guinea

Sulawesi

Solomon Islands

Java

Sumatra

Timor

Vanuatu

Fiji

AUSTRALIA

New Caledonia

New Zealand

Tasmania

ISLAND LOCATIONS

The map shows that most of the world's large islands lie close to the continents. Islands lying further out to sea, are usually smaller, although there are plenty of small islands around coasts, too. There are also islands in lakes and inland seas, and in rivers, too.

Continent or island?

Although Australia and Antarctica are surrounded by sea, they are continents, not islands. This is because they are parts of moving sections of the Earth's crust.

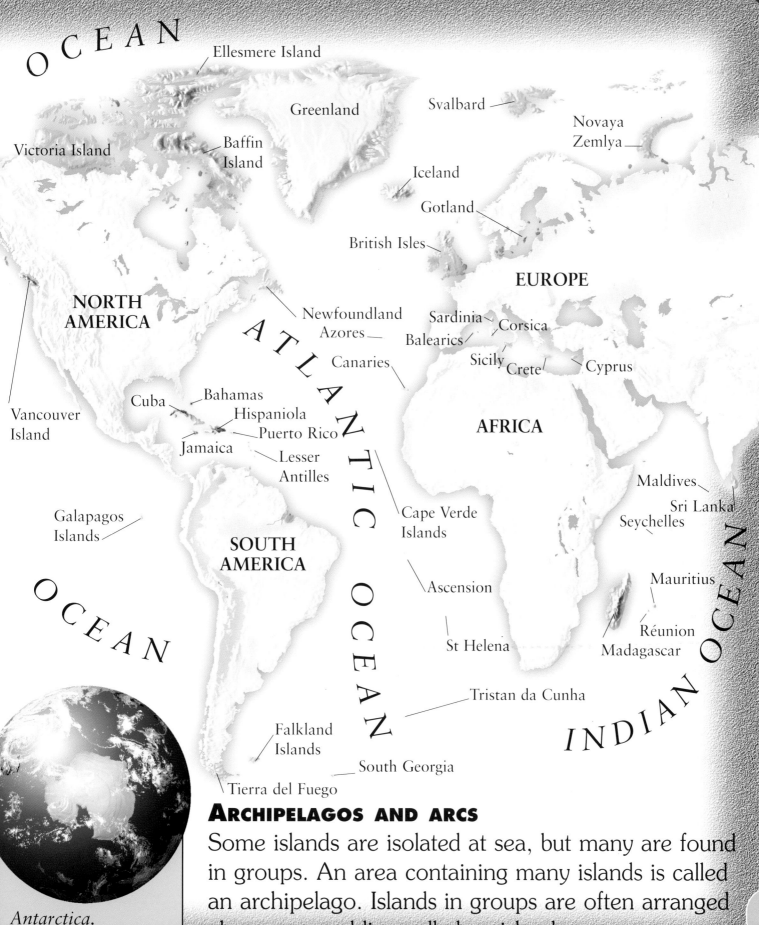

OCEAN

Ellesmere Island

Greenland

Svalbard

Novaya
Zemlya

Victoria Island

Baffin
Island

Iceland

Gotland

British Isles

EUROPE

NORTH
AMERICA

ATLANTIC

Newfoundland
Azores

Sardinia
Balearics

Corsica

Sicily
Crete

Cyprus

Canaries

Cuba
Jamaica

Bahamas
Hispaniola
Puerto Rico
Lesser
Antilles

AFRICA

Vancouver
Island

Maldives

Sri Lanka
Seychelles

Galapagos
Islands

Cape Verde
Islands

SOUTH
AMERICA

Mauritius

OCEAN

OCEAN

Ascension

Réunion
Madagascar

St Helena

INDIAN OCEAN

Tristan da Cunha

Falkland
Islands

South Georgia

Tierra del Fuego

Antarctica.

ARCHIPELAGOS AND ARCS

Some islands are isolated at sea, but many are found
in groups. An area containing many islands is called
an archipelago. Islands in groups are often arranged
along a curved line called an island arc.

7

WHAT IS AN ISLAND?

Very simply, an island is a piece of land that is completely surrounded by water. There is no official limit on how big an island can be, although Australia is considered to be a continent rather than an island.

Numerous small islands in a river delta in Madagascar. The delta is formed from sediment washed down the river.

New Zealand is made up of two large oceanic islands formed by volcanic activity. Many of its volcanoes are still active.

ISLAND TYPES

There are two main types of island – continental islands and oceanic islands. Continental islands are islands that lie on the edge of a continent and form part of that continent. Oceanic islands are islands out in the oceans that are not part of continents. The land that forms them rises up from the ocean floor.

Most isolated islands

The small group of volcanic islands that make up Tristan da Cunha are the most isolated islands on Earth. They lie in the South Atlantic, midway between South America and southern Africa. Only a few hundred people live on the biggest of the islands.

Tristan da Cunha.

Andros Island is the largest island in the huge archipelago that makes up the Bahamas.

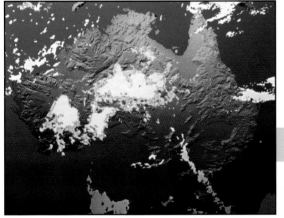

Australia and its surrounding islands make up the continent of Australasia.

MOVING CONTINENTS

The Earth's solid crust is made up of several huge parts called tectonic plates that float on the liquid rocks underneath. Two hundred million years ago, there was one supercontinent, Pangaea. Gradual movement of the tectonic plates has created the continents we know today.

200 million years ago   180 million years ago   60 million years ago   Today

9

Continental islands range from tiny offshore islands to enormous islands that make up a substantial part of a continent, such as Greenland in North America and New Guinea in Australasia.

CONTINENTAL EDGES

At a continent's edge, the Earth's surface drops steeply to the ocean floor down a slope called a continental shelf. Shallow areas of sea before the shelf often contain continental islands.

Glaciers are remnants of the huge ice sheet that melted at the end of the last ice age.

FORMING CONTINENTAL ISLANDS

Continental islands are formed when the sea floods a section of coastal land, cutting off the high ground. The diagram on the right shows a section of hilly coast with a peninsula. When the sea floods the low-lying land, the hilltops on the peninsula are cut off and create new islands off the coast, separated from the mainland by a shallow sea.

Whole peninsula is part of mainland.

MAINLAND AT TIME OF LOW SEA LEVEL

The British Isles are part of the European continent. They are cut off by the shallow North Sea.

SEA LEVELS RISE.

RISES AND FALLS

Continental islands are formed when coastal land is flooded. This happens either when the sea level rises, or the land falls. Many of today's coastal islands were formed about 10,000 years ago at the end of the last ice age, when a huge ice sheet melted and sea levels rose. Some coastal islands, such as the British Isles, were formed when sections of the land fell below sea level because of huge movements in the Earth's crust.

The Orkney Islands north of Scotland were created by sea-level rises 10,000 years ago.

Sea floods coast, leaving peninsula hills as islands.

Tidal islands

Some islands very close to the coast, such as St Michael's Mount in Cornwall, England, are only islands temporarily. At low tide, they are linked to the mainland by exposed rocks and sand. At high tide, the link is submerged and they become islands again.

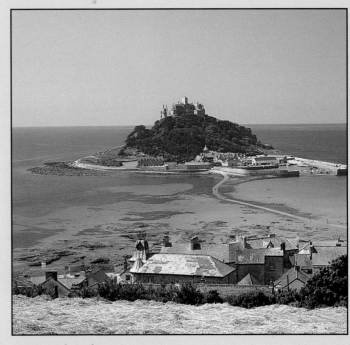

St Michael's Mount.

Isolated islands far out in the world's oceans are called oceanic islands. Their land is not part of any continent, but is formed by the summits of towering mountains rising up from the ocean floor.

UNDERSEA VOLCANOES

Where magma leaks through a crack in the seabed, it builds underwater volcanoes. Erupting volcanoes eventually break the surface, and form new oceanic islands.

Hawaii is the tip of a massive active volcano. Lava flows slowly to the sea, making Hawaii larger.

HOT SPOT ISLAND ARCS

Some oceanic islands form over 'hot spots' in the Earth. At a hot spot, magma breaks through the thin oceanic crust, building a volcano that may grow to form a volcanic island. As the tectonic plate over the hot spot slowly moves, the hot spot stays still, punching new holes in the crust and forming new volcanoes. Over millions of years a line of hot spot volcanoes can create an island arc.

Older volcanoes

Crust moves

Mantle

Youngest volcano forms over hot spot.

Hot spot

Iceland sits on top of the mid-Atlantic spreading ridge. The two halves of the island are slowly moving apart.

BOUNDARY ACTIVITY

Many oceanic islands form above the boundaries between tectonic plates. At a spreading ridge, where plates move slowly apart, magma moves up to fill the gap, often forming islands. Volcanic islands also form above boundaries called subduction zones, where one plate slides under another deep beneath the ocean.

Krakatoa

In 1883, a devastating volcanic eruption blew away two-thirds of the island of Krakatoa in Indonesia. The explosion caused a giant tidal wave (tsunami), that drowned thousands of people on nearby islands.

Krakatoa is still an active volcano.

13

The Pacific Ocean and the Indian Ocean are dotted with thousands of beautiful coral islands, such as the Maldives. The water around these islands teems with life.

A coral atoll in the Truk Palau islands in the Pacific Ocean. The lagoon is a dark blue colour.

CORAL REEFS AND ISLANDS

A coral reef is a ridge of coral that forms in shallow, tropical seas. Waves erode (wear away) the coral, forming sand that piles up on the reef, to make a coral island. Tropical plants take root in the sand, helping to bind it together.

The Pacific island of Bora Bora is made up of a central volcanic peak surrounded by numerous coral reefs and islands.

Coral polyps

Coral is made from the skeletons of millions of tiny animals called polyps. Different polyps form colourful corals in the shape of branches, fans and ridges.

A coral reef.

ATOLLS AND LAGOONS

Coral reefs can form offshore (where they are called barrier reefs), or close to shore (where they are called fringing reefs).

An atoll is a ring of coral reefs and islands surrounding a central lake, called a lagoon. Because of the way the reef grows, the land of a coral island is never more than a few metres above sea level.

CORAL ATOLL FORMATION

1 A coral atoll begins to form when an underwater volcano begins to grow over a hot spot on the ocean floor. 2 The volcano continues to erupt until it breaks the surface, creating a new island.

3 A fringing coral reef grows in the shallow water around the island, and the volcano becomes extinct. 4 The cone of the volcano begins to erode away, but the coral continues to grow. 5 Eventually the volcano cone disappears completely, leaving behind a coral atoll with a central lagoon.

3

4

5

The islands of the world experience the whole range of climates (patterns of weather), from polar to tropical. Some islands are so large that the climate in the centre is quite different from the climate on the coast.

CLOUDS AND RAIN

The seas that surround islands have a huge effect on island climates. In hot climates, it can rain almost every day on an island, as heat from the land encourages clouds to form in the moist air over the ocean.

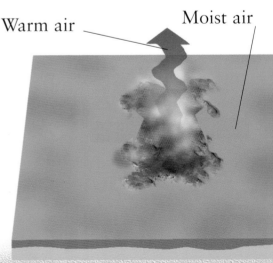

Palm trees on Bermuda being thrown about in hurricane winds, which can reach speeds of 250 kilometres per hour.

Ocean currents

Island climates are often affected by ocean currents, which bring warm or cold water to different parts of the globe. The Gulf Stream carries warm water from the Caribbean to islands in the North Atlantic, such as the British Isles. This makes their climates milder than other islands so far to the north.

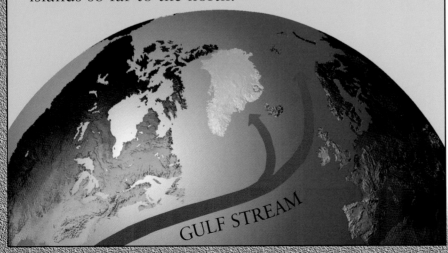

GULF STREAM

NIGHT AND DAY

The Sun warms the land and the land warms the air above it. The air rises, creating clouds and onshore breezes.

Warm air   Moist air

DAY TIME

ISLAND STORMS

Islands out in the oceans are unprotected from storms that bring destructive winds and flooding. Islands in the Caribbean are in the path of hurricanes sweeping in from the Atlantic, and the Philippines are in the path of Pacific typhoons.

The tight swirl of a hurricane system over the Atlantic Ocean, photographed from space. The hole in the middle is called the eye.

Islanders in Indonesia take cover from a seasonal storm.

After sunset the land cools, taking heat from the air above. This air sinks, creating offshore breezes, and the clouds scatter.

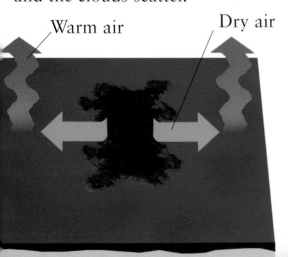

Warm air

Dry air

NIGHT TIME

Clouds forming over St Kitts in the Caribbean.

17

AN ISLAND IS BORN

Every so often, a new island appears. The most spectacular births are those of volcanic islands, which burst unexpectedly out of the oceans. New islands are also formed by growing corals, and by the action of rivers and tides.

BIRTH OF SURTSEY

In 1963, smoke and steam suddenly started to rise from the sea off the south coast of Iceland. The island of Surtsey was born. Just four days later it was half a kilometre across. Surtsey stopped erupting in 1967, and was soon colonized by plants, birds and insects.

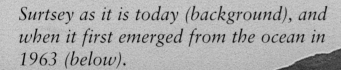

Surtsey as it is today (background), and when it first emerged from the ocean in 1963 (below).

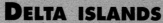

Islands in the delta of the River Betsiboka, Madagascar. The land is fertile, but farming can be hazardous because of floods.

DELTA ISLANDS

A delta is a fan-shaped piece of land, cut through by channels. It forms where a river dumps sediment at its mouth. When a river floods, the channels often change position, destroying some islands and creating others.

VOLCANIC BIRTH

Volcanic islands, such as Surtsey, form when magma from beneath the Earth's crust breaks out, forming a volcano. The eruption gradually builds up a cone-shaped mountain of lava. When the volcano breaks the surface, layers of ash are deposited as well as lava.

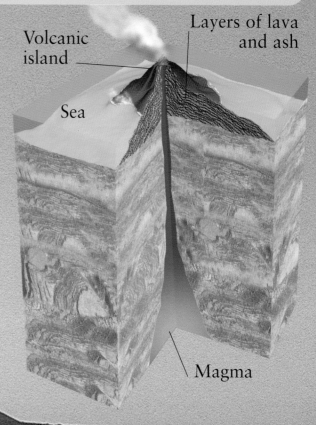

Layers of lava and ash

Volcanic island

Sea

Magma

Death of an island

Atlantis was a legendary island that sank into the sea thousands of years ago. Some people believe that Atlantis actually existed and was a volcanic island destroyed by catastrophic eruptions. One possible site for Atlantis is the volcanic island of Santorini, in Greece, which is the remains of a larger island.

Santorini in the Greek islands.

19

Lying between the North Atlantic and the Arctic Ocean is a vast island, permanently covered in a thick sheet of ice. This is Greenland, the largest island in the world. Despite the fact that Greenland sits almost completely inside the Arctic Circle, about 50,000 people live around its coasts.

Walking on Greenland's ice sheet. On average the ice is 1.5 kilometres thick, but reaches twice this thickness in the centre of the island.

GREENLAND

Ellesmere Island

Baffin Bay

Baffin Island

GREENLAND

Greenland is part of the continent of North America. It measures 2,655 kilometres from north to south, and 1,290 kilometres from east to west. It is almost two-thirds the size of Australia.

Main map

THE ICE SHEET

A thick sheet of ice covers 85 per cent of Greenland. The ice sheet never melts, and the summits of Greenland's mountains poke up through it. There are numerous islands and fjords around Greenland's coasts. Glaciers flow down from the ice sheet into the fjords, filling the fjords with icebergs. Most of the coast is free from ice and snow.

Viking settlements

Greenland was named by the Vikings, who first settled there in about AD 985, led by Eric the Red. The Viking colony grew to about 3,000 people before the climate cooled in the 1400s and made it impossible to stay.

The remains of Brattahild, Greenland.

Inuit people wear seal-skin clothes and travel by dog sled.

PEOPLE OF GREENLAND

Most Greenlanders live in towns on the island's south-west coast. They are descended from Inuit and European settlers. A few people live traditionally by hunting, but most work in the island's fishing industry.

All of Greenland's settlements are on the coast.

ISLAND WILDLIFE

The nature of an island means that is it is separated from other areas of land. This isolation has meant that some unique animals and plants have evolved on islands.

Plant seeds are often washed up on the shore.

1 Wind carries seeds and insects.

2 Small animals and plants carried by ocean currents on flotsam and jetsam.

Orang-utans are under threat of extinction because their rainforest habitat on Borneo and Sumatra is being destroyed.

Komodo dragon.

ISLAND EVOLUTION

Where a piece of land is cut off to create an island, species often begin to evolve differently from animals on the mainland. The Komodo dragon is a lizard that evolved only on some islands in Indonesia.

COLONIZING AN ISLAND

When new volcanic islands or coral atolls are formed, the land is lifeless. But plants and animals soon begin to colonize them. Birds and insects fly there, and plant seeds are carried by the waves or wind. Large land animals cannot colonize oceanic islands.

3 Bird droppings deliver seeds.

4 Humans bring animals and plants which may destroy an island's delicate balance.

Giant plants

The coco-de-mer, a palm tree from the Seychelles, has the largest seeds of any plant. They drift across the sea and take root when they land.

A sprouting coco-de-mer.

FLIGHTLESS BIRDS

Many species of birds only fly to escape predators. On remote islands where there are no predators, flightless birds, such as the kiwi of New Zealand, have evolved. Unfortunately, many flightless birds, such as the kakapo, also of New Zealand, are hunted by domestic animals introduced by man.

Frigate birds fly to Aldabra in the Indian Ocean to breed. A lack of predators means that the birds' eggs and young are safe.

The Galapagos Islands in the Pacific Ocean, and Madagascar in the Indian Ocean are treasure chests of nature. On these islands there are strange species of animals that are not found anywhere else in the world.

THE GALAPAGOS

Among the amazing animals that have evolved on the Galapagos Islands are giant tortoises, and the marine iguana, the only species of lizard that lives in the sea. Fourteen species of finches also live on the islands, each slightly different from the others.

THE GALAPAGOS

Isla Pinta

Isla Marchena

Isla San Salvador

Isla Fernanda

Isla Isabela

Isla Baltra

Isla Pinzon

Isla Eden

Isla Santa Marcea

Main map

Charles Darwin, the British naturalist, visited the Galapagos Islands in 1835. Studying the unique animals, including the finches, helped him to develop his theory that animals evolve by natural selection.

The marine iguana feeds on algae in the sea off the Galapagos Islands. It is the only lizard to do this.

Isla Santa Fé

Isla San Christobal

I S L A N D S

Isla Española

Madagascar is the fourth largest island in the world. It split away from the rest of Africa about 100 million years ago.

MADAGASCAR

Most of the animals and plants in Madagascar are unique. The island's most famous animals are its lemurs. All the species of lemur are now under threat because their forest habitats have been destroyed.

Island visitors

Wildlife enthusiasts visit islands such as the Galapagos and Madagascar to see the unique wildlife for themselves. The money they spend helps to set up conservation programmes for endangered species.

Tourists encounter giant tortoises in the Galapagos.

A ring-tailed lemur. The lemurs use their striking tails for signalling to each other.

For thousands of years people have moved to the world's islands in search of new land for farming and settlement, or to explore a new way of life.

ISLAND LIFESTYLES

On remote islands, such as those in Polynesia in the South Pacific, people live traditional lifestyles. They survive by fishing, raising animals and growing crops for food. The Polynesians colonized these islands tens of thousands of years ago, travelling vast distances across the Pacific Ocean in canoes. Isolated island life is not always easy, and many young people leave to find work on the mainland.

On islands such as Hong Kong and Singapore, people live modern lifestyles in bustling cities. Hong Kong has become a major centre for international commerce.

Polynesians are excellent sailors. They fish and travel in narrow dug-out canoes.

Borneo is the world's third largest island. Most of its inhabitants are people called Dayaks. Many still practise traditional crafts, including weaving beautiful cotton cloth.

Connecting islands

For island countries, such as Japan, that have developed into commercial and industrial centres, fast transport links are vital. Road and rail bridges and tunnels are built to link islands to each other and to the mainland.

Most of the six million people of New Guinea live in villages. In lowland villages, thatched houses are built on stilts to keep them dry in the rainy climate.

Kyoto Multiple Bridge, Japan.

The islands of the world boast a wealth of resources, from abundant sea life and rich farming land to rare minerals and geothermal energy. Beautiful island scenery and beaches also draw tourists by the millions.

The fertile, black volcanic soil and the mild Atlantic climate of the Canary Islands make the islands ideal for growing grape vines for wine production.

ISLAND PRODUCTS

The main industries on islands are fishing and mariculture, such as fish farming and pearl farming. Islanders fish for food and also export their catches. Some island climates are good for growing crops and rearing animals. The rocks of volcanic islands often contain useful minerals such as gold and gemstones, which are mined and sent to the mainland or to other countries.

Atomic atolls

The remoteness of the Pacific islands is not always in their favour. The people of Bikini Atoll were moved to other islands so that the USA could test nuclear weapons there. The USA is now paying for the islanders to move back to their native areas.

Nuclear bomb test at Bikini Atoll.

Leatherback turtles lay their eggs on tropical beaches. But their nesting sites are being badly disrupted by visiting tourists.

Coral is collected from reefs for building material and to make souvenirs for tourists. Coral reef destruction destroys the precious reef habitat that has taken thousands of years to grow.

ISLAND TOURISM

Island landscapes, from the white sandy beaches of the Caribbean, to the rugged volcanic features of Iceland, make islands popular tourist destinations. The economies of many islands, such as the Balearics, off Spain, depend on tourism. But thousands of tourists in new hotels also create problems, such as water shortages and pollution.

The perfect island hideaway. This is Beachcomber Island, one of the hundreds of islands in Fiji, with exclusive luxury holiday lodgings.

THE LARGEST ISLAND	Greenland is the largest island in the world. It has a total area of 2,175,600 square kilometres. That's almost ten times the size of Great Britain. More than 1,800,000 square kilometres of Greenland are buried under ice. Greenland's wiggly coastline is about 5,800 kilometres long.
THE BEST OF THE REST	The rest of the top ten islands by area (given in square kilometres) are: New Guinea (808,510), Borneo (757,050), Madagascar (594,180), Baffin Island (508,000), Sumatra (473,600), Honshu (230,500), Great Britain (229,880), Ellesmere Island (212,690) and Victoria Island (212,200).
THE LARGEST BY CONTINENT	The largest islands on each continent, except Antarctica, are: North America: Greenland (2,175,600 square kilometres); South America: Tierra del Fuego (47,000 square kilometres); Europe: Great Britain (229,870 square kilometres); Africa: Madagascar (594,180 square kilometres); Asia: Borneo (757,050 square kilometres); Australasia: New Guinea (808,510 square kilometres).
TALLEST ISLAND	The tallest island, measured from the ocean floor, is Hawaii. Its highest volcano, now extinct, is Mauna Kea. Its summit is 4,205 metres above sea level, but an astonishing 9,200 metres above the ocean floor. That's makes Hawaii taller than Mount Everest. Kilauea, also on Hawaii, is the world's most active volcano.
THE COUNTRY WITH MOST ISLANDS	Indonesia is made up of about 13,600 islands. Many of these are less than a kilometre across, but Indonesia includes parts of the huge islands of New Guinea and Borneo, making a total area about the size of Greenland. About 6,000 of the islands are inhabited.
THE MOST ISOLATED INHABITED ISLAND	Tristan da Cunha, with a population of just 300 people, lies in the South Atlantic, about 2,600 kilometres from Cape Town, South Africa and about 3,000 kilometres from Rio de Janeiro, Brazil, making it the most isolated inhabited island in the world.
THE MOST AT RISK	The average level of the land in the Maldives, in the Indian Ocean, is only 2.5 metres above the sea. If global warming causes the sea-level to rise, the Maldives will be the first islands to disappear under the waves.

GLOSSARY

archipelago
An area containing many islands, or a large group of islands.

fjord
A long, narrow, steep-sided valley that runs inland from the sea, filled with seawater.

global warming
The gradual increase in the average temperature of the Earth's atmosphere. It is probably caused by the burning of fossil fuels.

ice age
A period when the Earth's temperature dropped and thick sheets of ice and glaciers extended south, covering much of North America, Europe and Asia.

lava
The name given to magma when it leaves a volcano.

magma
Hot, molten rock that lies just underneath the Earth's solid crust.

peninsula
A long strip of land, almost surrounded by water, but connected to the mainland at one end.

Polynesia
A huge group of islands in the Pacific, stretching from New Zealand to Easter Island.

spreading ridge
A boundary between two tectonic plates, where the plates gradually move apart.

subduction zone
A boundary between two tectonic plates, where one plate slides underneath the other.

tectonic plate
A huge section of the Earth's crust. The crust is broken into several tectonic plates.

tsunami
An ocean wave caused by an earthquake, volcanic eruption or landslide.